Scientists and Primary Schools

a practical guide

Brenda Keogh and Stuart Naylor

Sponsored by

Supported by

Millgate House Publishers

Millgate House Publishers
Millgate House
30 Mill Hill Lane
Sandbach
Cheshire
CW11 4PN

First published in Great Britain in 1996

ISBN 0 9527506 0 0

British Library Cataloguing in Publication Data.
A catalogue record for this book is available from the British Library

Graphic design and typesetting by Kathryn Stawpert, The Manchester Metropolitan University

Artwork by Ged Mitchell, The Village Gallery, Wheelock, Sandbach

Covers Printed by The Printing House, Wistaston Road Business Centre, Crewe

Document printed by The Manchester Metropolitan University

Contents

Our grateful thanks are due to all the friends and colleagues who have been involved in creating this book:

Monica Winstanley and **Tracey Reader** at the Biotechnology and Biological Sciences Research Council, for their confidence that this book was a worthwhile venture and for sponsoring its publication.

Brian Gamble at the British Association for the Advancement of Science, for his trust in our judgement and for the support provided for our work.

Kathryn Stawpert at The Manchester Metropolitan University, for translating our typed notes into a work of art.

Ged Mitchell at the Village Gallery, Wheelock, for his illustrations and for bringing our cartoons to life.

Alan Goodwin at The Manchester Metropolitan University, for his constructive criticism, ideas and wise advice.

The many **colleagues**, **teachers**, **students** and **children** who have provided the inspiration and the misunderstandings which have resulted in the cartoons and who have given such helpful feedback on how the cartoons might be improved.

We are grateful to **Plymton St Mary C of E Infants' School**, nr Plymouth for permission to reproduce the fair test for clothes pegs from ASE Primary Science (1988) No 25.

Introduction

The idea for this book was conceived at a conference for scientists and teachers organised by BBSRC in October 1995. We presented the outcomes of our research into how cartoons could be used as the starting point for learning in science to the conference, and the response of the scientists present was very positive. Our desire to provide support for scientists was welcomed by Monica Winstanley, Head of Public Relations at BBSRC. This book is the direct result of her support.

The book is addressed to "scientists". We are using this as a shorthand term, including in it technologists, medical practitioners, engineers, astronomers and all the other science-related professions. We hope that you will bear with us if you feel that our description lacks accuracy.

Our intention is to provide direct support to "scientists" working with primary schools. We recognise that this is a valuable role for scientists to take on, with benefits to the scientist, to the school and to the wider community. We also recognise that it is a challenging role and that it is easy for misunderstanding and misinterpretation to reduce the value of the scientist's involvement. By describing some of the factors to consider when working with schools we hope to give you greater confidence in that role. By offering a series of cartoons and other starting points for investigation we hope to make your work with schools more effective.

It would not be feasible to mention every possible safety hazard which could arise. Instead there are two general guidelines. Firstly, discuss the details of every activity with the teacher in advance. While children are on school premises teachers have a direct responsibility for their safety, so the teacher needs to be kept fully informed at all times. Secondly, obtain a copy of "Be Safe" from Association for Science Education (see page 58). This is the most concise and useful guide to safety in primary schools which is available.

Scientists and primary schools

The nature of science in primary schools

Times have changed since many of us were at school. Science is now an important part of the curriculum for all children of primary school age, including children in nurseries. It is one of the three core subjects of the National Curriculum which applies to all children in state schools in England and Wales.

A typical class might spend the equivalent of one afternoon each week on science-related activities. Usually the science will take place in ordinary classrooms, on ordinary desks or tables. Most schools will have a limited amount of "specialist" science equipment, such as thermometers and magnifiers, but not necessarily enough for each child in a whole class to use the equipment at the same time.

Perhaps the most important aspect of the National Curriculum for science is the emphasis which is placed on investigation. Although there may be times when the children follow detailed instructions, for much of the time they are likely to be discussing, questioning and investigating ideas for themselves. Science will generally be an active, hands-on, exploratory experience. This is the context in which you will be working with schools.

The value of scientists working with primary schools

All scientists have a lot to offer to primary schools. Even if you know that your area of specialist expertise is not included in the National Curriculum for primary school children, you may be able to find some overlaps and examples of the application of ideas in everyday life. What may surprise you is that schools may be far more interested in your role as a professional scientist and your general scientific awareness than they are in your specialism.

Helping children to investigate for themselves need not require specialist knowledge; creative imagination, coupled with understanding of general principles, is usually more important.

Many (though not all) primary schools will welcome and value your involvement. Through you they can get access to a community of scientists, to up to date knowledge, to fresh insights and new ideas. They can get a real sense of the creativity and personal satisfaction involved in science, technology and engineering. They will be able to get direct support in an area of the curriculum where many primary teachers lack confidence.

You can also benefit from your involvement with the school. It will give you useful insights into how primary schools function, and although it is demanding there is tremendous satisfaction to be gained from working with young children and helping them to become enthusiastic and confident about science.

Your contact with a school may be a one-off occasion. This will have some value. However, the most important benefits to you, to the school and to the wider community will come from the mutual understanding which develops over longer periods of time. A long term relationship with a school will be better for you, better for the school and ultimately much more satisfying for everyone involved.

The role of scientists working with primary schools

If you see your value in working with primary schools as being limited to passing on your specialist expertise then your role may be very restricted. By contrast, when you see yourself as valuable because of your general scientific awareness then you can take on a greater range of roles. Most of your contact is likely to be working with the children in school, but other forms of contact will also be useful.

Some of the most significant ways of working with primary schools are outlined below.

Acting as a role model

You are a living illustration of what scientists do and of what science is about. You can't avoid this role even if you would like to!

Illustrating the value of science

You can be an important link between the school and the world outside. Your general awareness of scientific ideas, of how scientific ideas are applied and of their importance in everyday life give you a clear sense of the value of science. Some of this can be shared with the children.

Responding to the children's enquiries

This will be especially valuable when the children are researching a topic and need more expert guidance than is available in school. Although you may not be able to answer all the questions that the children raise, you will be able to answer most of them from your general scientific background.

Complementing the approach used by the teacher

Clearly you will need to find out what approach the teacher normally uses. It is likely to involve the children in activity rather than just listening. It is likely to involve leaving the children with questions as well as answers.

Supporting the teacher as well as the children

Many teachers will feel intimidated by your expertise, just as you may feel intimidated by theirs! They should be able to tell you how you can use your expertise to make a valuable contribution to the school. It is important to work on the school's agenda, not your own. Helping the teachers to develop their understanding of scientific ideas may be an additional benefit that emerges.

Being involved in science outside lesson time (eg through a science club)

This is less demanding than being involved with the whole class, since there will usually be fewer children and they will have chosen to be there. It can help to reinforce the point that science is an interesting and enjoyable activity that people choose to engage in. It can also be a good way to get to know the children, teachers and possibly parents in less formal settings.

Supporting special events

Some schools organise science fairs, science competitions or science open days. These are demanding events to organise, so advice, encouragement and practical support will normally be welcomed. As with science clubs, this can be a small scale and lower risk level of involvement.

Providing support for individual projects

Children working individually on projects will generally need some support. You may be able to help with ideas, science background, resources and encouragement. Useful support can be offered by letter or by telephone as well as in school.

Providing resources

Not all scientists are in the position of having resources to give away. However, out of date equipment (such as microscopes and balances) and everyday items (such as plastic dropping pipettes or dishes) can make a big difference to schools which are short of basic resources.

Welcoming children into your work environment

This may not always be feasible. Where it is this will be a valuable opportunity for the children to extend their understanding of how scientists work.

Mistakes that scientists may make

You may already have a significant involvement in one or more primary schools. If so, then involvement as a parent, as a governor, as a voluntary classroom assistant or as a member of a Parents Association should have given you considerable insight into the reality of school life.

Alternatively you may not have had much opportunity to get involved. Your expectations of what goes on in primary schools might be based on vague recollections of your own school days and coloured by the selective accounts of schools which are presented by the media. You may not have an accurate picture of what schools are like now. In these circumstances there is a risk that what you feel able to offer is not what the school feels it needs. Everybody then ends up disappointed and valuable time may be wasted.

Fortunately some of the obvious mistakes that scientists may make can be predicted and avoided.

Expectations

There is no point in talking with the children if they don't understand what you are saying. Judging the language, the vocabulary and the pace of learning of young children is not always easy. This is something to check out with the teacher in advance. Similarly, if you are working with the children in school, it will be useful to check the children's likely concentration span, their willingness to contribute and their likely behaviour. Whatever you do, don't bring lots of exciting things into school and expect the children to stay calm and quiet!

You will also need to check what level of involvement the school is hoping for. If things go well and you find yourself spending more time with the school, don't forget that you might only be seeing a part of the teacher's job. Science is not the teacher's only concern. They also have to enthuse the children in every other subject, plan lessons, assess the children, go to meetings, liaise with parents, write policy documents, prepare for school inspections and so on. Expectations on both sides have to be realistic.

Approaches

Clearly the approach that you use may vary according to the age and capabilities of the children you are working with. Younger children will usually need more hands-on experience, more concrete examples, more references to their experience and a more "dramatic" style of presentation - though these are useful principles for older children's learning as well.

More generally there will be problems if you just tell the children lots of facts and don't attempt to relate new ideas to their experience. You may also find that the children are less likely to listen to you if you don't appear interested in listening to what they have to say. You will be seen as an expert by the children and the teachers, so it isn't necessary to appear to know all the answers. Realising that even the experts do not know everything is an important part of the children's learning.

Relationships

There is no question that you have a lot to offer to the school. However if you do not feel that you have anything to learn in return then your relationship will be less productive. Being willing to give but unwilling to take may simply intimidate teachers, damage their confidence and weaken your relationship with them.

Primary teachers are professionals in their own field. They have a wealth of professional expertise which is different from yours. A relationship based on mutual respect will allow you to make the most of what you have to offer. Including the teacher at the planning stage, drawing on their expertise whilst in school and remembering at all times that you are a guest in their classrooms will make your contributions even more valuable.

Your relationship with the children is no less important. You cannot expect young children to respond only to your subject expertise. Their responses will be on a much more personal level than that. They will make judgements about what science is and what scientists do from what they see in you. Sometimes it may not be possible to avoid fitting the stereotyped picture of a scientist. Whether we like it or not some of us are white, male, middle-aged, balding, with beards and glasses! If these bits of the stereotype fit then it is all the more important to avoid confirming some of the other aspects of the stereotype - cold, aloof, impersonal and insensitive. The picture of the scientist that you leave the children with is as important as the pictures of science that you create.

Finally, don't forget to relax and enjoy the experience - and don't give up if things sometimes go wrong.

Getting investigations started

An essential feature of science in primary schools is that children will be investigating. Through observing, exploring and investigating more systematically they will gradually develop the skills of working scientifically. Sometimes discussion and mental involvement will be enough to develop the children's ideas, but usually some kind of hands-on experience will be necessary.

By working with the children during lessons, in science clubs or on individual or group projects you can support this process. By providing possible starting points for investigations you can open up new areas for the children to explore.

Whatever you ask the children to do you will need to keep some important points in mind.

- Activities need to be **purposeful** to the children. The context for the activity therefore needs to be carefully chosen.
- Activities need to be **accessible**. The concepts involved and the language used must not be too complex for the children concerned.
- Activities need to **actively involve** the children. They should be able to contribute their own ideas at some point and not simply work with other people's ideas.
- Activities need to have the **"sell it to them"** factor. They need to capture the children's attention and motivate them to find out more.

Some of the ways in which you can capture the children's attention and interest are described below, with suggestions as to how investigations may emerge from these starting points.

Everyday objects

Everyday objects can be used as starting points for investigation by highlighting some of their features and asking questions about them. Even though the object may be familiar the children will realise that they can find out more about it.

Questions such as which paper towel is best at mopping up water or which shape of cup is most stable when it is full of liquid are examples of investigations based on everyday objects. Figure 1 shows an example of an investigation into which clothes peg is the strongest. Getting each child to bring a peg from home ensures that the investigation is linked to the children's direct personal experience.

FIGURE 1

FROM ASE PRIMARY SCIENCE (1988) NO 25

we used a dolls cup and a buckit and Some Sand and 3 pegs

One child put a peg on the handle of a plastic beach bucket and held it up in the air.

A second filled the cup with sand -

we Smoothed off the Sand with our Fingers

then poured it into the bucket. "*If you pour it in gently you can get more in before the bucket falls.*"

They counted the cups of sand they loaded into the bucket before it fell off the peg.

Everyday events

Starting investigations from everyday events will also allow you to build on the children's personal experience and to relate new ideas to those which are already understood. Questions need to be raised which help the children to realise that their understanding of the event is incomplete.

Examples of suitable everyday events could be water evaporating from a dish of water or ice melting. Where does the water go to as the dish dries up, how quickly does it evaporate and what can speed it up or slow it down are the kinds of questions that the children could investigate.

Interesting or unusual objects

In these examples the aim is to provide an object which will interest the children and hold their attention. They will naturally want to play with the object and explore what they can do with it. From here it is only a small step to investigating more systematically by focussing on specific questions about the object.

Objects which can provide interesting starting points include paper helicopters (see ASE Primary Science 1982, No 7), ice balloons (balloons filled with water and then frozen - more details in Primary Science Review 1987, No 3) or balancing toys like the one shown in Figure 2. This can be constructed simply from wooden skewers and plasticine. It will fascinate the children in the way that it can balance at almost any angle. The children will be able to investigate how the amount and the position of the plasticine affect the way that the model balances.

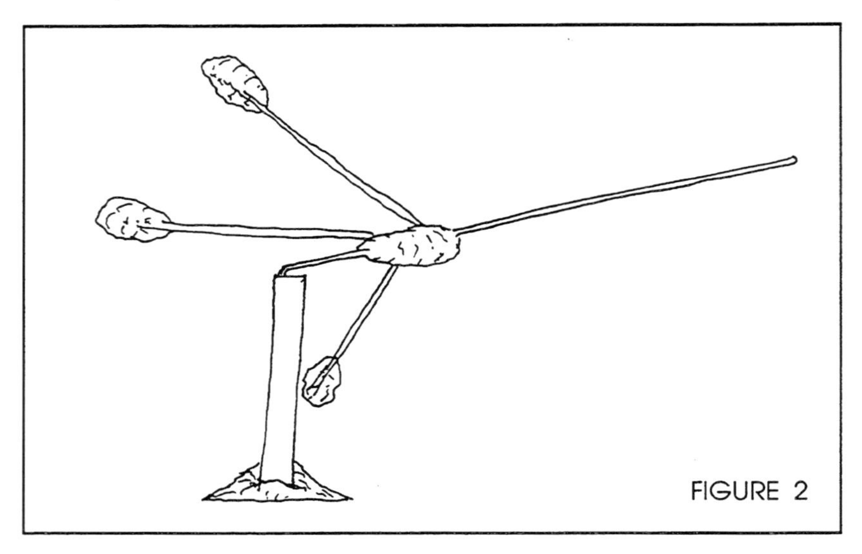

FIGURE 2

Interesting or unusual situations

As with the interesting or unusual objects, the aim is to provide experience of a situation that will excite the children's curiosity and hold their attention. The children's interest will lead to their asking questions and exploring the situation; questioning and exploring can then lead them into more systematic investigation.

Examples of interesting situations include the simple circuit shown in Figure 3, where pressing the switch will make the lamp go out. In Figure 4 the flattened piece of plasticine can be made to float and the flattened piece of aluminium foil made to sink. However screwing up the ball of plasticine always makes it sink while screwing up the ball of aluminium foil always makes it float!

FIGURE 3

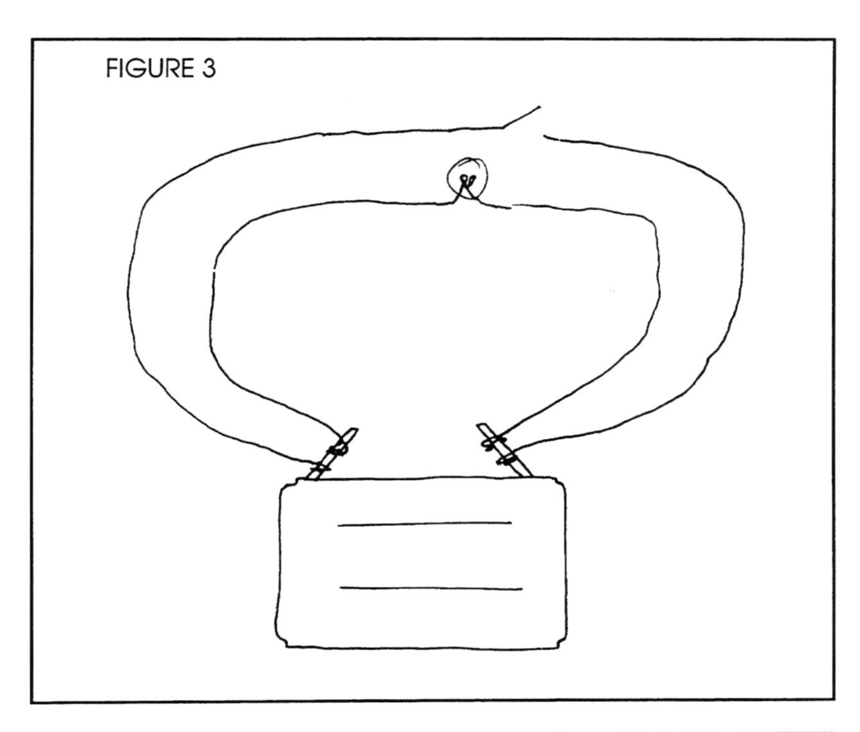

FIGURE 4

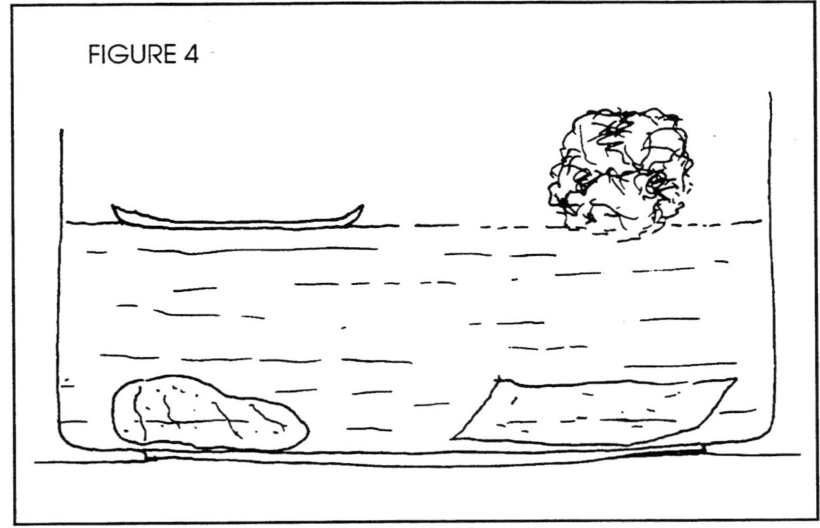

Observing the natural environment

The natural environment provides a rich resource for children's investigations. Close observation of any plant, animal or biological system will not only provide information but will also readily generate further questions. Not all the questions that children raise can be investigated. However a useful part of their learning in science is to realise that some questions are more productive and more easily investigated than others.

Examples of starting points for investigations include food preferences in animals (eg which flowers do the bees seem to prefer?), habitat preferences (eg do the woodlice seem to prefer dry or damp conditions?), plant distribution (eg which plants seem to survive best where people walk?) and seed germination (eg do the biggest seeds grow fastest?). Ideally most of the questions will come from the children themselves.

Sorting and classifying

Inviting the children to decide how to sort things into groups and then to classify them using the categories that they have chosen can be a useful starting point for investigation. Classifying objects requires the children to make judgements. Frequently they will find that further investigation is necessary before these judgements can be made.

For example, inviting the children to classify materials as solids, liquids or gases can lead into investigations about viscosity. How thick can a liquid be and still be runny? Is it possible to have a runny solid? Can we find a simple measure of runniness?

Similarly classifying living things as animals or plants can lead to the children having to find out more about their characteristics. Do all animals have fur? Do all animals move? Are there any plants that move? Do all plants have roots? A set of cards with pictures of living things can generate lively discussion and provide a basis for the children wanting to investigate and find out more. Figure 5 shows some possibilities. A fascinating example of how ordinary children responded to this activity is given in Primary Science Review 1992, No 21.

FIGURE 5

Posing a problem

Posing a problem to the children can be a useful starting point. The problem involves them in a situation, and they will need to investigate the situation further in order to solve the problem. There is no shortage of problems suitable for this purpose. *Primary Science* and *Primary Science Review* (published by Association for Science Education - see page 58) are rich sources of problems that can be used in this way, so you do not need to feel that you have to invent all the problems yourself.

Some examples of suitable problems could be: -

- making a waterproof hat for Paddington Bear
- deciding on the best place to fix the school fire bell (ASE Primary Science 1989, No 28)
- making a 2-minute timer (ASE Primary Science 1988, No 25)
- can we revive a bunch of wilted flowers? (ASE Primary Science 1987, No 22)
- finding out which are the strongest conkers (ASE Primary Science 1984, No 14)
- deciding which type of egg boxes are best (ASE Primary Science 1981, No 6)

Cartoons

Cartoon-style drawings which illustrate possible areas of uncertainty for the children can be a powerful means of providing starting points for investigation. Section 3 includes 30 cartoons covering a wide range of science concepts which you may be able to use with the children.

Each cartoon presents the children with alternative viewpoints on some scientific concept. They invite the children to share their views and to explore which of the alternatives presented is more likely to be correct. In some cases discussion may be sufficient for the children to develop their ideas, but usually the children will be desperate to start investigating to find out more about the situation presented in the cartoon! Through investigation they will be able to develop their ideas further.

The cartoons are not necessarily designed to have a single right answer. In many cases the only possible answer is "It depends on . . . ." This is a realistic perspective on science for the children to develop. Most of the cartoons can lead to a wide range of possible investigations, with the children's initial ideas determining which investigation will be most relevant to them.

The cartoons can be copied and given to the children or they could be redrawn much larger and used as a focus for a group or class discussion. Having small groups of children working on a cartoon together is usually the most effective arrangement. However they can also be a useful starting point for an individual project or as an extra challenge to some of the children.

Other possibilities

There are lots of other possible starting points for investigations. You will almost certainly have ideas that you can take in from your workplace which can be made accessible to the children. There will also be resources in or around the school which can provide starting points, such as story books, workschemes and the school environment. Other starting points for investigations can come from the local or national news, from the weather, from a trip round a supermarket or from a visit by the children to your workplace where this is feasible. A creative imagination and a willingness to allow the children to generate their own questions and ideas will usually be sufficient.

Cartoons as starting points for science

Living things and life processes

Plant in water

The issue in this cartoon is where plants get their food. Does giving them sugar from an external source make any difference to their growth? The children can investigate this with various types of plants and they can use a number of different measures of whether the plant has grown, such as length of stem, size of leaves, number of new leaves, etc.

Seeds and cotton wool

Sometimes incidental factors can seem very important to the children. If they germinate seeds on cotton wool they may not have a clear idea of what the cotton wool (or paper towel or tissue or sponge or . . .) is for. The children can investigate the various possibilities put forward in the cartoon by adjusting each of the factors separately.

New growth

Identifying where new growth occurs in plants is not obvious to children. The cartoon invites them to observe plants closely and to investigate where the new growth takes place. Part of the children's investigation will involve deciding how they will be able to spot the new growth when it happens. They may come up with ideas such as making detailed drawings, taking photographs, marking the plant at regular intervals or looking for differences in appearance between new plant tissue and old.

Runners

Many biological investigations do not allow "experiments" to be set up to observe the effect of changing one of the variables. Sometimes we can only examine a biological system carefully and attempt to identify relationships which may be present. This is the case with the runners. The children can look for a correlation between how fast children can run and the other factors suggested in the cartoon. They would need to collect data systematically in order to identify any possible link. The genetic link between children and their parents is mentioned as one possibility. It is important to treat any questions about the children's family backgrounds with great sensitivity and not to make any assumptions about their backgrounds.

Size of seeds

The children may suggest other factors which may affect the rate of growth, such as the shape of the seed, how wrinkly it is and how thick the outer coat is. Each of the different factors involved can be isolated and investigated separately, including the genetic link between seeds and their parent plants. The children will have different ideas about what growing faster means - the time it takes for germination to begin, the rate of growth of the root or the rate of vertical growth of the shoot are all possibilities.

Is it an animal?

Often children will only identify mammals as animals, with rapid movement on four legs usually being their most significant feature. The issue is whether being an animal excludes being in another group (such as the insects) or whether animal is an inclusive term, so that the bird can be both a bird and an animal. This cartoon invites the children to rethink their existing definition and to look for a consistent meaning for the word.

plants in water

seeds and cotton wool

I wonder where the new bits will be as the plant grows
At the bottom of the leaves
At the top of the leaves
All the way through the leaves
In the middle of the leaves
It just grows new leaves

new growth

runners

Who can run the fastest?
The children with the biggest feet
The children with the longest legs
The children who can take the biggest breaths
The children who are the lightest
The children whose parents are runners

size of seeds

is it an animal?

Light

Cards and shadows

The issue in this cartoon is whether shadows are affected by the colour or thickness of the object. Children will not necessarily have a well-formed idea of what a shadow is. They may confuse shadows with reflections and this will influence their expectations of what affects the shadow. They can investigate the situation as it is shown in the cartoon. Investigating a range of cards of different thicknesses and colours will be relatively straightforward. Access to a light meter or light sensor could be useful for the older children. Investigations may be complicated by other light sources which are present and by reflection from the cards, especially if some of the cards have a glossy surface.

Shadow stick

The issue here is the position of the sun rather than anything to do with the stick. Observing the stick's shadow at different times of day will reveal that the shadow changes its length and position as the sun moves. This can be a lead in to further investigations about how the sun moves and how this changes during the year. Making a simple sundial could be a useful follow up activity. Other investigations could focus on the relationship between the light source and the object, where the intensity of the source or the distance and the angle between the source and object can be altered.

Bending mirror

Precisely how curved mirrors reflect light is complex. Children are unlikely to understand the concepts involved in any depth. However it is possible for them to investigate this situation systematically and to gain a better understanding of the various ways in which images can be formed. By using a flexible plastic mirror they can investigate the differences between reflections in concave and convex mirrors and the difference that the degree of curvature makes to the reflected image. Setting up a fairground hall of mirrors will be an enjoyable way for the children to apply their understanding.

Headlights at night

There is an obvious possibility of confusion between the distance that the light travels and how well a light source illuminates an object. Children may well see these as synonymous but any investigation needs to separate them. The children can investigate whether light is received by an observer in conditions of different background lighting. For example, they could investigate whether an observer can see a small torch beam at a distance at different times during the day when background illumination varies. Separate investigations could be carried out into how well a torch beam illuminates an object in conditions of different background lighting. A useful follow up could be to investigate how to make objects highly visible (eg road signs) and how to make them invisible (eg animals being camouflaged to avoid predators).

cards and shadows

shadow stick

bending mirror

headlights at night

Materials

Snowman

What exactly is the effect of the coat on the snowman? Is the coat generating heat and melting the snowman, or is it simply an insulator which will prevent the snowman from melting? The children can investigate the situation shown in the cartoon using a simple model to help develop their ideas. Real snow is ideal, but ice cubes inside a glove or sock could be a reasonable model, or a "snowman" could be made by freezing water in the top half of a plastic mineral water bottle.

Ice pops

All of the predictions in this cartoon can be directly investigated by the children. Some of them are likely to think that aluminium foil is an insulator; that cotton wool makes things warmer; that water will keep the ice pop cold; and that things will stay frozen inside a refrigerator. Their ideas about the thermal properties of materials can be challenged and developed through their investigations.

Ice cream

Although the children will have experience of condensation they are unlikely to have well-formed ideas about where the condensed water comes from. The cartoon invites them to consider and investigate a number of possibilities, and they may well think of other possibilities themselves. Some of the children may see the air as the source of the condensed water as the least likely possibility. Investigations such as this help to lay the foundation for later work on the structure of matter and conservation of mass.

Is it a solid?

The children will have intuitive ideas about what they mean by a solid. Whether they can apply their criteria in a consistent way in making judgements is more doubtful. The cartoon provides an opportunity for them to rethink their definitions and to make more systematic judgements. Introducing more challenging materials such as sand or dough is probably best left until after their ideas about solids are reasonably well developed.

Sandcastles

The distinction between melting and dissolving is a common area of confusion for children. They can clarify the meaning they attach to both of these terms by investigating the situation shown in the cartoon. Most primary schools will have a sand tray which can be used to model the effect of the tide on sand castles. Observation of other changes in materials, such as melting chocolate or dissolving sugar, will be a useful complement to their investigation.

Sugar in tea

This cartoon invites the children to reverse the familiar process of dissolving. It also challenges the children's ideas about what happens to the sugar in the tea - does it disappear completely as it dissolves or can it be recovered from the tea? The children can investigate the possibilities shown in the cartoon as well as well as other possibilities that they might suggest. Salt is a useful alternative, since it can be separated more easily from water than the sugar. Other means of separating materials and other examples to illustrate conservation of mass would be useful ways to follow up this investigation.

Don't put the coat on the snowman - it will melt him

I think it will keep him cold and stop him melting

snowman

ice pops

How can we stop the ice-pops melting?
Put them in a dish of water
Wrap them up in cotton wool
Put them in the fridge
Put them on the window sill
Wrap them in foil

Why does the ice cream tub get wet if I leave it out of the freezer?
It was already wet when you took it out of the freezer
The water must have leaked out of the container
Perhaps the freezer isn't working properly
The water has come from the air
The drops might not be water
ICE CREAM
ice cream

is it a solid?

All of these objects are solid
The elastic band isn't - it can change its shape
The pencil isn't - it will float in water
The ball bearings aren't - they don't stay where you put them
The feather isn't - it is too light

sandcastles

sugar in tea

How could we get sugar back out of the tea
I don't think that you can get the sugar back
Boil it up so that the water evaporates
Put it in the freezer
Put it through a strainer
Stir it really hard

Sound

Big ears

The issue in this cartoon is how much difference the external ears make to our hearing. The cartoon suggests that the ability to judge direction and the apparent volume of a sound may be affected by the size of the ear. Each of these can be investigated separately. The children can make themselves large ears out of card for their investigation and can close their eyes while they listen to sounds being made by other children. You may need to find a quiet area for this to be manageable. The children's investigations can lead in to follow up work which looks at the ears of animals in real life and relates the size of ears to life style.

Guitar string

The guitar string provides a manageable way to investigate how the thickness of a string affects the nature of the sound produced. The children will quickly realise that the length and tension in the string also make a difference. These need to be kept constant as far as possible during the children's investigations. Their results should be applicable to any stretched string, so they should be able to make their own instrument from elastic bands stretched over a hollow container. They may also be able to apply general principles, such as the size of the object affects the pitch of the note produced, to other musical instruments. Some of these other instruments could be modelled, using bottles or tubes to model organ pipes, plant pots to model bells or cymbals, and so on.

String telephone

This cartoon identifies several factors which might affect the way that sound is transmitted. All of them can be investigated by the children with very limited apparatus. Their investigations should begin to develop their ideas on sound transmission and sound amplification. This can lead in to further investigations on how well sound is transmitted by different materials, how well sound is transferred from one medium to another and how the shape of an object can affect sound transmission.

Glass to the wall

This cartoon illustrates a possible area of confusion between glass as a material and a glass as an object. Children will tend to expect that the characteristics of each will be similar and not recognise the additional properties that a glass has by virtue of its shape and size. Investigating how different shaped objects can transmit or amplify sound and what kinds of materials make good sound insulators will help to develop the children's ideas. Older children may find that a sound sensor is a useful aid to their investigation. The children's investigations can easily be related to real-life situations where sound transmission or insulation are important, such as soundproofing a noisy teenager's bedroom or ensuring that the fire bell can be heard all the way round the school.

Will I hear better if I make my ears bigger?
You will be able to tell where the sounds are coming from better
You will only hear sounds from the front
The sounds will seem louder
It won't make any difference

big ears

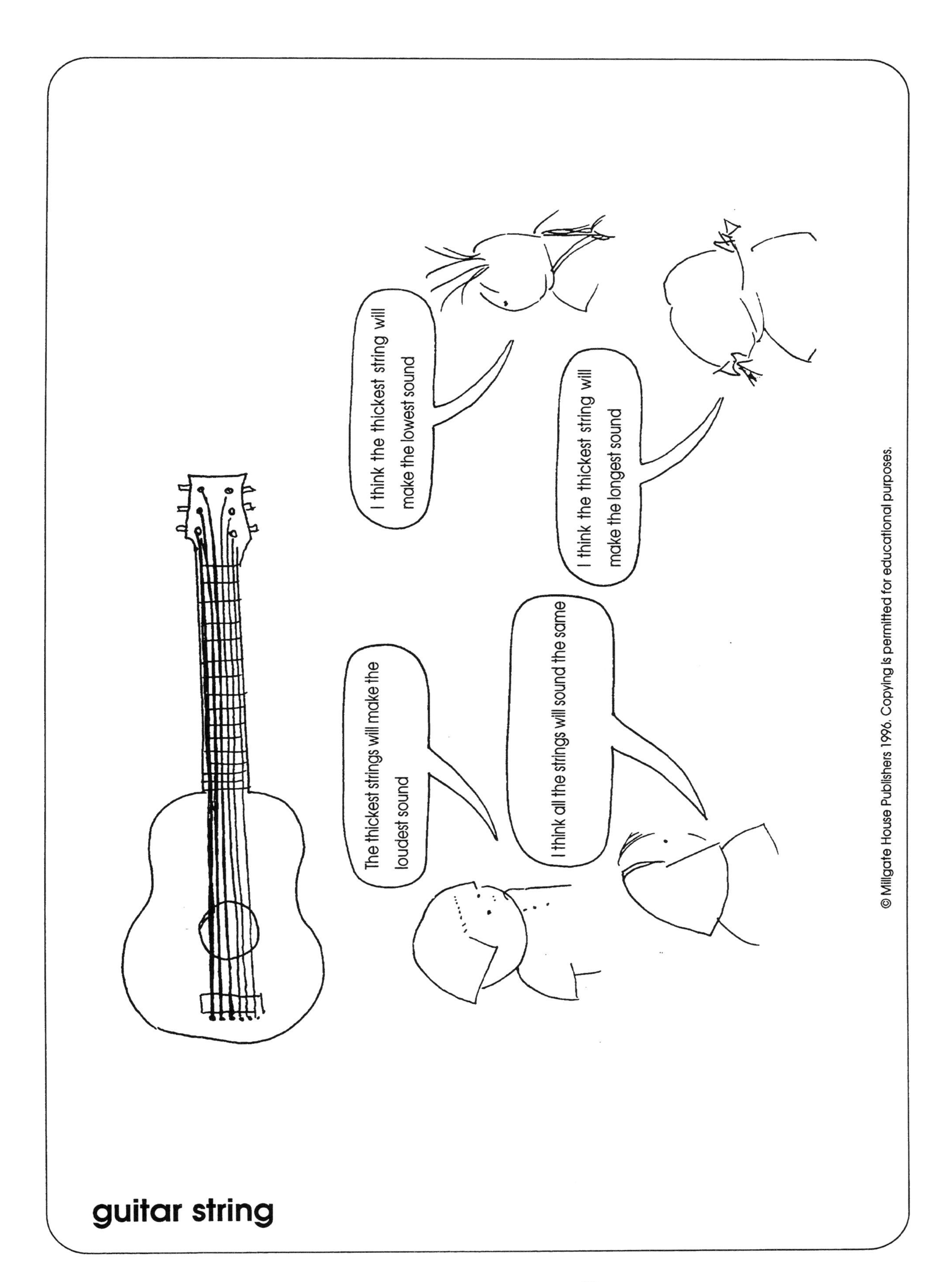

guitar string

string telephone

glass on wall

Living things and their environment

Plants use up soil

In this cartoon the issue is whether plants obtain their food from the soil, with obvious consequences for the soil level. The children can investigate plants growing in varying amounts of soil, including no soil at all, and measure any apparent changes in the soil level. The suggestion that some changes may be too small to notice introduces a degree of tentativeness into their thinking. Whether the plants use up soil can lead in to the more fundamental question of how plants do obtain their food.

Rubbish bin

This cartoon links recycling and rotting. If something can be recycled does that also mean that it will rot? The children can investigate whether a range of items will decompose over a period of time and they will need to take into account the conditions which are likely to lead to decomposition. It will be useful to draw a distinction between organic ("made from something that used to be a part of a living thing") and inorganic in their investigation. Care should be taken with sharp edges on the cans and the risk of breakage to the glass; some schools will not allow children to handle glass.

Leaves in woodland

The children can set up investigations to explore whether leaves decompose or whether they are eaten by a variety of invertebrates. In this cartoon the possibilities are not mutually exclusive. Either possibility can be a useful introduction to understanding how natural cycles work. The children will need to consider what the conditions are like when leaves fall off the trees and they will need to find some way of preventing leaves from blowing away so that invertebrates can eat them.

Frogs and mud

The possibilities in this cartoon are not mutually exclusive. It would be extremely difficult for the children to carry out a practical investigation into this situation. Instead they will need to research different aspects of the frog's life history and make judgements based on the information available to them. They may appreciate some pointers, such as finding out what herons eat or finding out what eats worms, slugs and beetles.

Fir trees

The children may be able to identify other situations where plants do not grow under trees. If they can then this may be a better focus for a practical investigation. It is possible for them to investigate some of the factors mentioned by growing plants in pots under the trees and by growing plants in soil taken from under the trees. It is unlikely that any investigation will provide conclusive proof since a number of these factors are likely to interact in real life. The children do not need to have a single definite answer to the problem posed in the cartoon for their ideas to develop.

Weeds and vegetables

Careful observation of a vegetable patch will reveal that it is only certain weeds that grow faster than certain vegetables. The children can carry out a number of investigations to compare the rate of growth of different weeds and vegetables in various soils. By observing weeds in a real vegetable patch the children may discover that many weeds are opportunist species with short life cycles and that this is one reason why some weeds are successful.

plants use up soil

rubbish bin

Will everything in the bin rot if we put it on the compost heap?
The pencil shavings will rot, but they will take a longer time
The cans will rot - they can be recycled
And the plastic bags will rot - they can be recycled too
Glass can be recycled so that must rot as well
The paper will not rot

leaves in woodland

frogs and mud

There are no plants growing under the fir trees because they take all the water from the other plants
They take all the light from the other plants
They don't leave enough space for the other plants
They poison all the other plants
They take all the food from other plants

fir trees

weeds and vegetables

Forces

Dropping objects

This cartoon illustrates the question of whether the weight of an object affects the speed at which it falls. This can be investigated directly by the children. Other factors such as the size and shape of the falling objects can also be investigated. A useful follow on is for the children to investigate how quickly objects with a large surface area fall (eg parachutes). Other investigations can focus on the speed of falling of objects which are light for their size (eg polystyrene blocks) and on how to increase or decrease the speed of falling. These further investigations will help to develop their understanding of how the effect of air resistance can be significant in determining how quickly an object falls.

Bathroom scales

Young children will not necessarily separate weight from pressure. They may not realise that there is any distinction between weight and pressure or how they are connected. Their ideas can be developed by investigating the situation described in the cartoon, using bathroom scales as shown. They could investigate similar situations using blocks of wood pressing down onto a slab of plasticine. Measuring the depth of the indentation in the plasticine will allow them to investigate the effect of weight and pressure separately. Real life applications of these ideas can be seen in elephant's feet and stiletto heels!

Wood on ramp

Here the question of how friction affects sliding is important. The children will have developed intuitive ideas about friction and its effects from their everyday experience. This cartoon invites them to develop their ideas in a more systematic way. All of the factors shown in the cartoon can be investigated by the children. They can explore the effect of the size, shape and weight of the block on sliding; they can alter the size of the frictional force by using smoother or rougher surfaces and by using various lubricants. They may also identify other features as important, such as the angle or length of the ramp. In order to be meaningful they will need to keep all these other factors constant as they investigate one of the variables involved. This situation provides a useful opportunity to consolidate the children's ideas on how to carry out a fair test.

Boat in deep water

The issue in this cartoon is whether the depth of water affects the way that an object floats. Many children (and some adults) believe that it does. They do not necessarily relate their beliefs to the real life situation of passenger ships or oil tankers. The children can investigate the situation shown in the cartoon using tanks or bowls of water. Rectangular boats made from folded aluminium foil can provide an easy way to identify whether the boat floats differently in different depths of water. This may well be part of a more general investigation into floating and sinking, in which the effects of other factors such as shape, size, type of material and presence of air can be explored.

dropping objects

bathroom scales

The block of wood will slide better if it is heavier
It will slide better if it is bigger
It will slide better if it is wet
I can think of something else to try
It will slide better if it is smoother

wood on ramp

boat in deep water

Useful contacts

Support for working with primary schools can be obtained from a wide variety of organisations. Support may take the form of ideas for activities to use with children, resource materials and visual aids, sources of advice, meetings, local networks and so on. Some of the most useful sources are listed below.

Association for Science Education
College Lane,
Hatfield,
Herts. AL10 9AA
Tel. 01707 267411

This is the most significant professional association for people involved in science education. It produces a range of publications for primary schools, including Primary Science Review and Primary Science. The publication **Be Safe** is the standard text on safety for science in primary schools in many Local Education Authorities. ASE runs meetings at local, regional and national level, and access to networks of teachers and educators can be obtained through these meetings. The Annual Conference is also a superb opportunity to contact many of the other groups and organisations which may be valuable to you. You or your organisation can join ASE as a primary member; we strongly recommend that you do.

The Biotechnology and Biological Sciences Research Council (BBSRC)
Polaris House
North Star Avenue
Swindon SN2 IUH
Tel 01793 413302

The Biotechnology and Biological Sciences Research Council (BBSRC) is one of seven UK research councils funded principally through the Government's Office of Science and Technology. It promotes and supports basic, strategic and applied research in the biological and related sciences. The BBSRC supports a total of over 7000 research scientists, postgraduate students, and technical and support staff in universities and institutes throughout the UK.

As part of its programme of activities designed to promote public awareness, appreciation and understanding of science, the BBSRC works to support science education in schools. It encourages scientists to develop links with schools, and produces schools' publications, posters and interactive materials for classroom use by teachers and pupils. For the primary sector a new set of posters and teacher handbook "Learning about Life" has been produced jointly by BBSRC, the Medical Research Council and The Association of the British Pharmaceutical Industry. BBSRC sponsors several school-based and pupil oriented events. It runs a free "Science Club" for teachers which currently has about 2000 members in the secondary and primary sectors.

To find out more about the BBSRC schools' liaison activities please contact: **Tracey Reader,** Schools' Liaison Officer.

To find out more about BBSRC's programmes on public understanding of science please contact: **Dr Monica Winstanley**, Public Relations.

British Association for the Advancement of Science

Fortress House,
23 SavileRow,
London W1X 1AB

The BAAS offers direct support to primary schools through a network of regional coordinators. The main focus of the coordinators' work is to support Young Investigator science clubs. These offer opportunities to children to become more involved in science and to gain awards for their involvement. Bronze awards are given for a series of short activities, while silver and gold awards are given for extended projects. The BAAS also organises national events such as BAYSDAY, a science activity day for children, and Science & Technology Week, where the focus is more on the public understanding of science.

Support, resources and advice can also be obtained from:

* Research Councils, such as BBSRC
* Professional associations, such as the Institute of Physics
* Industrial associations, such as the Association of the British Pharmaceutical Industry
* Individual organisations, such as Unilever or British Telecom
* Museums and science centres, such as Curioxity in Oxford
* Conservation groups, such as The Wildlife Trusts

The simplest way to make contact with many of these groups is through the ASE Annual Conference or through the BAAS.